Cover illustration: After an agreement between the United States and the USSR regarding anti-ballistic missile defence systems, the USA activated its Safeguard system. This comprised the long-range Spartan missile and the close-range Sprint. Sprint (illustrated) was developed to destroy enemy ICBMs that evaded Spartan and threatened to hit their US targets, Sprint's nuclear warhead to be a 'last stand' defence against the incoming warheads. However, Safeguard missiles were deactivated in 1976 as an economy measure.

1. The Western Electric M-1 (later MIM-3A) Nike Ajax was not only the US Army's first supersonic anti-aircraft missile but the first such weapon to become operational in the United States, in December 1953. It was a two-stage command-guided missile, with a high-explosive warhead and range of 25 miles (40km). Some 15,000 were produced, being deployed by the US Army in Europe and the Far East, the National Guard in the US and by several European and other armies.

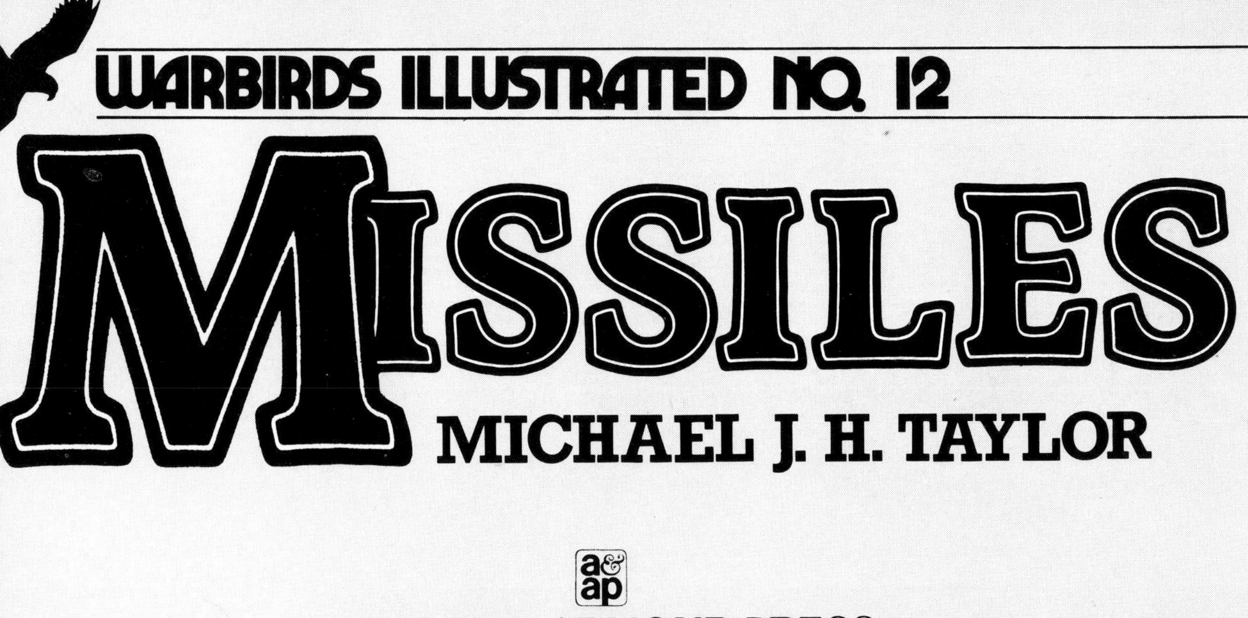

a&ap

ARMS AND ARMOUR PRESS
London—Melbourne—Harrisburg, Pa.

USMC
COBRA
USMC

Introduction

This book differs from all previous volumes in the Warbirds Series by dealing with *unmanned* flying weapons and their manned carriers, built, tested and deployed over four decades. All too often aircraft, ships and land vehicles have received world attention when first shown publicly, with little or no regard for the weapons fitted. But each and every military conflict has shown the carrier to be less important than the weapon, a classic case of the 'punch' not the glove. This volume attempts to redress the imbalance and in this endeavour includes photographs of historical, long-since deactivated missiles as well as modern missiles and missile carriers from many different nations. Some it can be claimed changed the world, others failed to be deployed at all through design failure or economic restraint, but the majority served or serve operationally.

Where possible, photographs have been grouped according to the subject's role, resulting in the 'old and new' being spread evenly throughout the volume. It has not been the intention to illustrate every type of missile currently deployed or ever built; this would be impossible in a volume of this size. Instead, illustrations have been selected on the basis of interest alone, with several sequences each showing a weapon and the result. Here, then, is an aspect of military hardware too often ignored.

Michael J. H. Taylor, 1982

Warbird 12: Missiles
Published in 1983 by
Arms and Armour Press, Lionel Leventhal Limited,
2-6 Hampstead High Street, London NW3 1QQ;
4-12 Tattersalls Lane, Melbourne, Victoria 3000, Australia;
Cameron and Kelker Streets, P.O. Box 1831, Harrisburg, Pennsylvania 17105, USA

British Library Cataloguing in Publication Data:
Taylor, Michael
Missiles. - (Warbirds illustrated; 12)
1. Guided missiles
I. Title II. Series
623.4′51 UG1310
ISBN 0-85368-571-1

Layout by Anthony A. Evans.
Printed in Great Britain by William Clowes, Beccles, Limited.

2. A US Marine creeps through the woods carrying a Cobra wire-guided anti-tank missile, 1959. Weighing just 22.5lb (10.25kg), but with a warhead then capable of destroying any known tank, Cobra was developed by MBB in Germany. The current version has its range extended to 6,560ft (2,000m) and is known as Cobra 2000.

▲3 ▼4

3. Of all the missiles designed and built in Germany during the Second World War, none is better remembered than the V 1. Correctly named the Fieseler Fi 103 or FZG-76, it was powered by an Argus 109-014 pulse-jet mounted simply over the winged fuselage. To Londoners, hit by more than 2,400 between June 1944 and March 1945, this *revenge weapon* was known as the doodlebug or buzz bomb. Its warhead comprised 1,874lb (850kg) of high explosive. Here, a V 1 is photographed at its vulnerable cruising altitude, with an RAF Spitfire closing in for the kill.
4. The German A4 rocket or V 2 *revenge weapon* was the ancestor of the modern ballistic missile, perfected by the legendary Wernher von Braun and others at the secret research establishment at Peenemünde. Much larger, six times heavier but with a warhead only slightly greater than that on the V 1, the hypersonic V 2 was first fired offensively (against Paris) in September 1944.
5. In an attempt to increase the V 2's limited range, two missiles were tested at Peenemünde fitted with sweptback wings. Both A4bs were unsuccessful.
6. An early French bombardment missile was the Sud Aviation SE 4200, powered by a ramjet in the fuselage and rail-launched using two solid-propellant boosters. Guidance was by radio command. Development started in 1950 and this, and the improved SE 4280, entered French service in limited numbers during that decade.

5▲ 6▼

▲7 ▼8

▼9

7. The Martin TM-61 Matador entered service in 1951. It was the first operational guided missile deployed by the USAF, equipping Tactical Missile Wings in Germany, Korea and Taiwan as well as in the United States. Powered by an Allison turbojet engine, it was launched from its transporter aided by a booster rocket motor. Such was its importance, the 1,000th example was delivered in 1957. The warhead could be nuclear or high explosive, and the just-subsonic missile had a range of 500 miles (805km). This TM-61C Matador of the 58th Tactical Missile Group is being set up at a launching area in Korea. (US Air Force)
8. A Matador thunders from its launcher under turbojet and booster motor power.
9. A Snark intercontinental missile blasts off from Cape Canaveral.
10-12. A sequence showing the nose-cone of a Snark missile separating from the expendable main airframe.
13. Airmen of Strategic Air Command make final checks on a Northrop SM-62 Snark before sending it hurtling along the South Atlantic test range. One of the first missiles with true intercontinental range to be deployed operationally outside the Soviet Union, it could carry a thermonuclear warhead over 6,300 miles (10,140km), the nose-cone separating after the cruise flight for final impact. Snark first entered service in 1959 with the 702nd Strategic Missile Wing at Presque Isle Air Force Base.

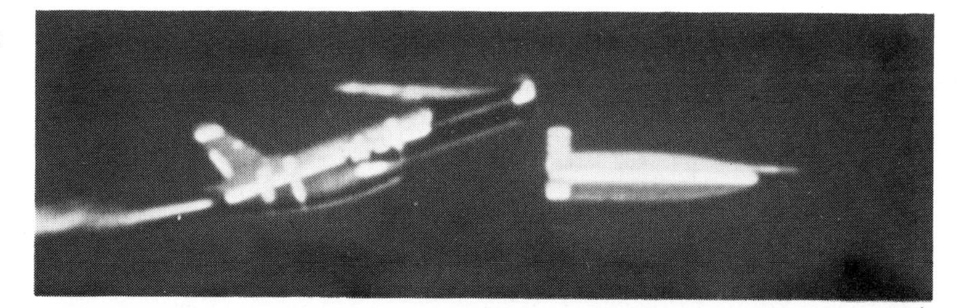

10▲

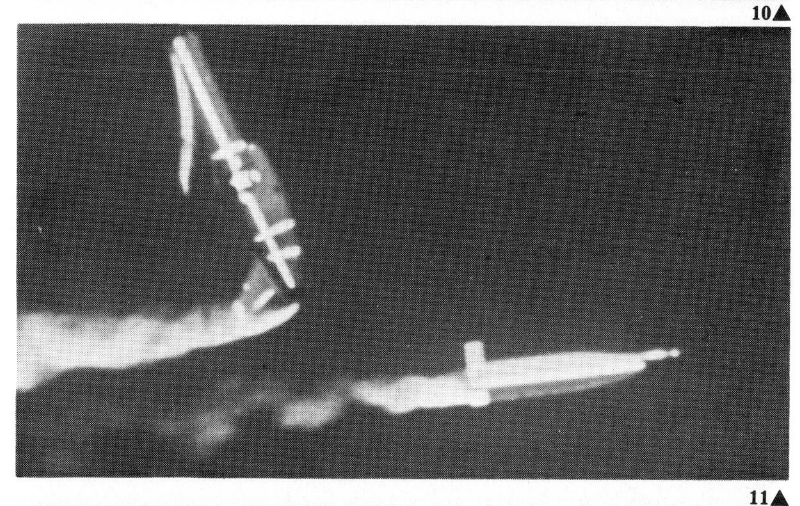

11▲

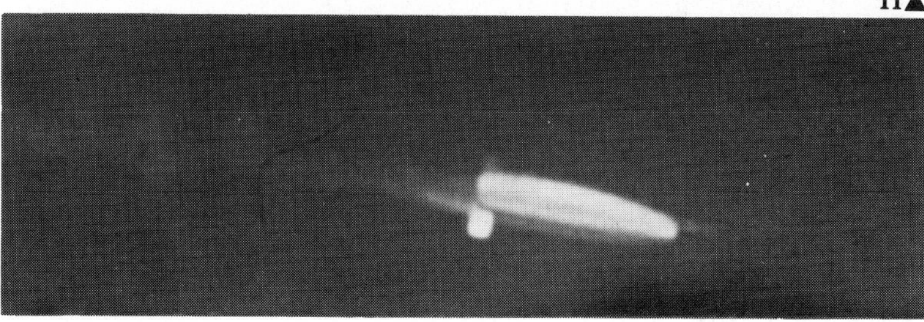

12▲ 13▼

▲14 ▼15

14. The Martin TM-76 Mace was basically a refined Matador with more than twice the range, as indicated by its original designation TM-61B. Nuclear armed, it entered production in 1958 and became operational with the USAFE's 38th Tactical Missile Wing the following year. Here, a Mace is being transported on a Goodyear 'Translauncher', pulled by a 'Teracruzer'.

15. The MGM-18A Lacrosse was an early field artillery missile, designed by the Cornell Aeronautical Laboratory and produced by Martin. The missile entered service with the 5th Missile Battalion, 41st Artillery Group on 22 July 1959. Launched from a US Army 2½-ton truck, it had a range of 20 miles (32km). Four US Army Lacrosse missile battalions were deployed at home, one in Japan and three in Europe.

16▲

16. The British English Electric Blue Water, a mobile selective-range field artillery missile, was not destined to become operational. A unit comprised a 3-ton launcher/transporter and a Land Rover carrying the ground computer. The missile was powered by a solid-propellant rocket motor with alternative nuclear or high-explosive warheads. Had the missile entered service, it would have replaced US-developed Corporals serving with the British Army and perhaps missiles of other NATO countries.

17. Development of the US Army Missile Command Honest John unguided field artillery rocket was started by Douglas in about 1950. It became operational eventually with the US Army, USMC and several NATO countries. However, it was quickly dropped by the USMC, as it proved difficult to move during amphibious operations. Carrying either a nuclear or high-explosive warhead, it could attain a speed of Mach 1.5 and had a range of 23 miles (37km). Here, a refined MGR-1B Honest John, which replaced the MGR-1A from 1960, is seen on its launcher. The missile could be prepared and fired by fewer than six men.

17▼

▲18 ▼19

18. NATO has used the reporting name 'Frog' for a series of Soviet field artillery missiles that have been operational in various forms since the 1950s. 'Frog' stands for 'Free Rocket Over Ground', indicating that they are unguided. Those shown here are 'Frog-5s', two-stage missiles, capable of a 34-mile (55km) range and carrying either nuclear or high-explosive warheads.

19. The Vought Corporation Lance guided artillery missile was developed as a modern replacement for Honest John and Sergeant, becoming operational in 1973. It can be fired from a tracked launcher or a lightweight wheeled launcher (as seen here). The range and speed are 75 miles (121km) and Mach 3 respectively.

20▲

21▲

20. A major milestone in American rocketry was the development of the Convair SM-65 Atlas, the West's first intercontinental ballistic missile. Unlike Snark, Atlas was a ballistic missile, being wingless and following a ballistic trajectory once powered flight had ended. Developed only after the production of a lightweight thermonuclear warhead became possible, the initial Atlas A test version with booster engines only and a dummy nose-cone was followed by the Atlas B, fitted with a proper nose-cone and booster and sustainer engines. Here, nearly ready for launch, is Atlas 12-B, which made the first full-range flight on 28 November 1958.

21. Atlas D launch. Fitted with a thermonuclear warhead in a blunt Mk 2 or lighter and pointed Mk 3 ablative type nose-cone (as shown), the Atlas D was declared operational in September 1959.

◄22 ▲23

▲24 ▼26

▲25 ▼27

22. The Martin Titan was the 'big boy' of the first-generation US ballistic missiles, in fact carrying the largest thermonuclear warhead ever fitted to a US missile (although not as large as some Soviet warheads). HGM-25A Titan I's first-stage rocket motor gave an incredible 300,000lb (136,100kg) of thrust, its second stage 80,000lb (36,290kg), allowing a speed at burn-out of nearly Mach 26. Titan I became fully operational in 1962, eventually equipping six squadrons, each with nine missiles. Here, a Titan I lifts off from Vandenberg Air Force Base, California. Two drawbacks of Titan I were the use of non-storable liquid propellants for its engines and the necessity of lifting it from its 160ft (49m) deep silo for launching.

23-27. LGM-25C Titan II overcame the shortcomings of Titan I, used more powerful engines and had a modernized ablative re-entry vehicle. It was so formidable that a total of 54 missiles in six squadrons were retained when the more modern Minuteman was deployed.

28. The Douglas Thor (USAF designation SM-75) was an intermediate-range nuclear ballistic missile or IRBM. The first successful launching took place in September 1957. Britain received 60 of these missiles, deploying them in four areas. This photograph shows one of No. 77 Squadron's Thors at Feltwell. RAF-operated Thors were deactivated in 1963, some of the returned missiles becoming space vehicle boosters in the United States.

29. Since the days when the United States could deploy by far the greatest ballistic force in the world, Soviet development has been continuous, well planned and successful. Today, Soviet ballistic missiles are the most numerous and formidable in all ranges. As can be seen from this 1960s photograph of a 'Scamp'/'Scapegoat' system, the mobility of Soviet IRBMs has meant that they can move undetected to new launching positions at will.

28▲ 29▼

▲30

▲31

30. Of the 1,054 intercontinental ballistic missiles currently deployed in the US, 1,000 are three-stage Boeing Aerospace Minuteman IIs and IIIs. Developed to replace first-generation missiles, Minuteman is considerably lighter and more accurate, but with a smaller airframe and warhead. Its solid-propellant motors allow virtually instant readiness, hence its name. All 800 LGM-30A Minuteman Is were operational by late 1965, but are no longer in use. LGM-30F Minuteman IIs have one re-entry vehicle each, but LGM-30G Minuteman IIIs use MIRV warheads, each with three re-entry vehicles to increase the chances of penetrating defences.

31. A Minuteman launch control centre under construction.

32. During 1974, trials were conducted to assess air-launching Minuteman missiles and so reduce the danger of first-strike annihilation of the land-bound force. Dragged by drogue parachutes from a Lockheed C-5A Galaxy, this test missile fell to about 8,000ft (2,400m) before ignition. Although successful, this concept went no further.

33. Test-firing a Regulus II, the 115,000lb (52,163kg)st Aerojet-General solid-propellant booster powering the missile from its launcher (see **103**).

34. During 1960-61 the USAF's Strategic Air Command deployed its first version of the McDonnell Aircraft Corporation ADM-20 Quail. Designed to be carried and launched by B-52 bombers, Quail was a decoy cruise missile. It was intended to fly in enemy airspace and confuse defences by giving off a radar signal similar to that of the bomber. Each B-52 could carry four Quails.

▼32

33▲ 34▼

▲35 ▼36

37▲

35. An early version of the Soviet medium-range anti-aircraft missile known to NATO as 'Guideline' was first seen in public in 1957. Since then it has become the most widely operated of all surface-to-air missiles originating in the USSR. The example illustrated is a 'Guideline Mk 2' operated by the Chinese People's Liberation Army.

36. Blowpipe, developed by Short Brothers, is deployed by British and other forces as a shoulder-fired, close-range, anti-aircraft missile, suitable for launching against oncoming and retreating aircraft. As with Redeye (see **134**), it uses a two-stage solid-propellant rocket motor, enabling the missile to be ejected from the launcher before the sustainer stage ignites, thus protecting the operator.

37. For more than twenty years, Soviet bombers have been able to carry various air-to-surface conventional and nuclear stand-off, tactical and strategic missiles. Of three early missiles seen at the 1961 Aviation Day Display, the first to be codenamed by NATO was the turbojet-powered AS-1 'Kennel'. These two AS-1s were carried by an Indonesian Air Force Tupolev Tu-16 'Badger', probably for an anti-shipping role. 'Kennels' are no longer operational.

38. One of the latest Soviet stand-off missiles is the AS-6 'Kingfish'. Rocket-powered, probably carrying a nuclear warhead, and with a cruising speed and range of Mach 3 and 135 miles (220km) respectively, it was first photographed by a Japan Air Self-Defence Force pilot in 1977. (JASDF)

38▼

BELL AIRCRAFT CORP
0613-B
RASCAL

▲39

15219

▲40 ▼41

H-36
H-34
H-32
H-30
U.S. AIR FORCE

39. Bell Aircraft began development of what became the GAM-63 Rascal in 1946. The first production Rascal was accepted by Strategic Air Command at Pinecastle Air Force Base, Florida, in October 1957, to become the USAF's first operational stand-off missile. Its cruising speed was Mach 1.6 and range 100 miles (161km).
40. The Boeing DB-47 Stratojet was the selected operational aircraft for the Rascal, one of which was carried on the starboard side of the fuselage.
41. XGAM-87A Skybolt was developed by Douglas after winning the USAF's ALBM (air-launched ballistic missile) competition against thirteen other contenders. A two-stage solid-propellant missile, capable of hypersonic speed and a range of 1,150 miles (1,850km), it was to have become operational on Boeing B-52s in 1964.
42. Another potential user of Skybolt was the RAF, which was to carry the missile on its Vulcan B Mk 2s. After the Skybolt was dropped by the US government, Britain also pulled out in favour of deploying Polaris submarines.
43. Blue Steel was developed by Hawker Siddeley Dynamics to provide RAF Vulcan and Victor bombers with the capability of attacking heavily defended targets from stand-off ranges. The warhead was thermo-nuclear. Operational development trials of Blue Steel were conducted on Vulcan B Mk 2s of No. 617 Squadron, RAF, from the summer of 1962. In February of the following year this squadron became fully operational.
44. After three Vulcan squadrons had become operational with Blue Steel at Scampton, it was the turn of Victor B Mk 2s based at Wittering, where No. 139 Squadron, RAF, became operational with the missile in September 1963.

42▲

43▲ 44▼

▲45

▲46 ▼47

45. The North American GAM-77 (later AGM-28) Hound Dog was the standard air-launched missile of the USAF's B-52 squadrons throughout the 1960s and the first half of the 1970s. It was a turbojet-powered nuclear stand-off missile with a speed of Mach 2 and range of 600 miles (965km). Two can be seen here below the wings of a refuelling B-52.
46. Shortly after Hound Dog became widely operational in August 1963, by which time 29 SAC wings used the missile, a successor was looked for that would prolong the useful life of the B-52 bomber. The answer came in the form of the Boeing SRAM (short-range attack missile), deployment of which began in 1972. SRAM is a Mach 3 nuclear missile with a range of 100 miles (160km) at high-altitude and 35 miles (55km) at low level. B-52G and H each carry twenty missiles, while FB-111As carry six.
47. Just becoming operational with the first few B-52Gs is the USAF's latest nuclear attack missile, the Boeing Aerospace AGM-86B ALCM (air-launched cruise missile). The ALCM, in its original AGM-86A form, was first air-launched under test on 5 March 1976 (as shown in the photograph). Development

48▲

then moved to the larger AGM-86B, which offers a range of 1,550 miles (2,500km). The missile's turbofan engine provides subsonic performance.

48. Probably the most famous air-to-air missile ever produced is the Sidewinder, developed by the US Naval Ordnance Test Station at China Lake (later US Naval Weapons Center) from 1948. The first firing took place in September 1953. Each missile has fewer than two dozen moving parts. Here, early AIM-9B Sidewinders are being fitted to a US Navy Grumman F9F-8 Cougar.

49. Over the years the capability of Sidewinder has been enhanced by continuous development, allowing the missile to remain in production today. Here, a British Aerospace Hawk has been modified to carry the AIM-9L Sidewinder, a widely used third-generation version with a speed of Mach 2.5 and range of 4.35 miles (7km).

50. The Soviet equivalent of the Sidewinder is the K-13A, known to NATO as AA-2 'Atoll' in its early infra-red guidance form and AA-2-2 'Advanced Atoll' with radar homing guidance. The multi-role versions of the Mikoyan MiG-21, represented here by this 'Fishbed-K', carry both versions of the missile.

49▲ 50▼

▲51

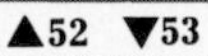

▲52 ▼53

51. In 1947 Hughes started the development of a semi-active radar homing air-to-air missile, the first production examples of which were delivered in 1954. Known as GAR-1 Falcons, these became standard missile armament on the Northrop F-89H Scorpion, three missiles being carried (with 21 rockets) in each wingtip pod and extended for launching.
52, 53. The Convair F-102 Delta Dagger also carried Falcons, initially in the GAR-1D (later redesignated AIM-4A) version with improved performance and manoeuvrability at high altitude. Here, three Falcons are fired from rails extended from the fuselage bay.

54▲

54. Like Sidewinder, the Raytheon AIM-7 Sparrow missile has become a standard weapon with NATO and other air forces: unlike Sidewinder, it is a heavy medium-range type with a higher speed and much larger high-explosive warhead. This USAF McDonnell Douglas F-4C Phantom II is seen on a photo mission in October 1965, armed with four Sparrow missiles. (US Air Force)

55. The modern missiles shown in photograph **54** are Sparrow IIIs, but the original Sparrow I was very different. Apart from having an earlier type of guidance system, it had a smaller body diameter, lower performance, and weighed only three-fifths that of the current AIM-7F. Four Sparrow Is are seen here in 1956, arming US Navy Chance Vought F7U-3M Cutlass.

55▼

▼56 ▲57

56. In an attempt to improve upon the already successful AIM-7E Sparrow, Hawker Siddeley Dynamics began a programme in the early 1970s to produce the Sky Flash, based on the US missile. This is a 'boost and coast' missile with a new semi-active radar homing head and other refinements. It is capable of Mach 4 and has a range of 31 miles (50km). Sky Flash will arm the RAF's Tornado F Mk 2 air defence variant when this becomes operational. Here, a prototype ADV carries four Sky Flash missiles under its fuselage.
57. The Hughes AIM-54 Phoenix long-range air-to-air missile was developed originally for the US Navy's General Dynamics F-111B, which did not enter service. The photograph shows an F-111B with a Phoenix during trials.
58. Phoenix got a new lease of life with the development of the Navy's Grumman F-14 Tomcat fighter. The Tomcat proved during trials that it could carry six Phoenix and attack several targets simultaneously. AIM-54A Phoenix became operational in 1974, carried first by Tomcats from USS *Enterprise*. An improved AIM-54C version has been developed. Speed and range of the AIM-54A are Mach 5 and 124 miles (200km) respectively.

58▲

▲59

▲60 ▼61

59. The British Fairey Fireflash appeared during the latter 1950s as a beam-riding air-to-air missile. With no sustainer motor, it was accelerated to speed by two jettisonable boosters, after which it coasted to the target. Missile jettisoning trials are being conducted here using the Boscombe Down Blower Tunnel.

60. After successful trials, which included the first target aircraft destroyed by a British missile, the Fireflash was ordered into production to arm Supermarine Swift F.Mk 7 in a training and tactics development role.

61. The British Firestreak air-to-air missile was designed by de Havilland Propellers but is best known as a Hawker Siddeley Dynamics type. An infra-red homing missile, it has the limitation of having to be launched from behind the target. It became standard armament on the Gloster Javelin (as seen, with protective covers on the missile), the Sea Vixen and Lightning interceptor.

62. This photograph shows clearly Firestreak's octagonal pointed nose, containing the guidance system, and the two rings of infra-red optics around the body which are used to make corrections for final interception. Speed and range of Firestreak are Mach 2+ and ¾ to 5 miles (1.2 to 8km) respectively.
63. The pursuit-course only firing capability of Firestreak led to the development of a Mark IV version, which subsequently became Red Top. More manoeuvrable, due to larger-area control surfaces, more powerful to allow Mach 3 performance, and with a heavier warhead, Red Top can be launched from any angle of interception. Here, a Hawker Siddeley Sea Vixen F(AW) Mk 2 displays Red Top.
64. Escorting away a Soviet Tupolev 'Bear' reconnaissance bomber during a 1974 NATO exercise is an RAF BAC Lightning of No. 23 Squadron. The Lightning carries two Red Top missiles, which are interchangeable with Firestreaks. (MoD)

62▲

63▲ 64▼

▲65

65. The medium-range Matra Super 530 has a maximum speed of between Mach 4 and 5 and is capable of destroying targets flying 29,530ft (9,000m) higher than the aircraft from which it was launched. First fired successfully from a Mirage F1 in 1976, it has been in use with the French Air Force and others since 1979.

66. The Mirage 2000, the French Air Force's latest interceptor and air superiority fighter, due for delivery from 1983, will also carry Super 530s. Illustrated is prototype 02 carrying eight 250kg (550lb) bombs and two Matra 550 Magic dogfight missiles. Widely used in France and exported, Magic has a speed of Mach 2 and a range of 3¾ to 6¼ miles (6 to 10km).

▼66

67▲ 68▼

67. A US Navy North American FJ Fury, carrying Martin Bullpup A missiles during trials in 1958. In the following year Bullpup A, based around a 250lb (115kg) bomb and with a solid propellant rocket motor, became operational. This Mach 1.8 tactical missile was steered to the ground target by the pilot using a hand switch, via which was passed radio command signals. Bullpup A was so successful that many US Navy and Air Force aircraft thereafter had it in their weaponry. The missile was also manufactured in Europe. A slightly longer-range Bullpup B version was also built, based around a 1,000lb (455kg) bomb but with an alternative nuclear warhead.
68. A Bullpup streaks from a Sikorsky HUS-1, to record the first ever launch of a radio-controlled missile from a helicopter, 1960. Bullpup A was then also the largest missile ever fired from a helicopter.

▲69

▲70 ▼71

69. The Fairchild A-10A Thunderbolt II is one of several aircraft capable of carrying the Hughes AGM-65 Maverick, a tactical air-to-surface missile, seen here as the outer weapons. The first production version was the AGM-65A, a TV-guided model, followed by the refined AGM-65B with 'scene magnification' to allow the pilot to 'lock on' to a target at greater range. Versions with imaging infra-red and laser seekers have also been produced.

70. The MBB Kormoran is a German-developed anti-shipping missile with a speed and range of Mach 0.95 and 23 miles (37km) respectively. Air launched from an aircraft flying at very low level, to avoid detection by the ship's radar, Kormoran then descends further to strike the target close to the waterline. First deployed on Marineflieger F-104G Starfighters (as seen), Kormoran will be used on Marineflieger and Italian Air Force Panavia Tornados.

71. Kormoran impacts on a target ship.

72▲

72. An RAF Buccaneer strike aircraft carrying three Martel missiles, the nearest in anti-radar form and the remaining two TV-guided. A French and British developed missile, Martel is carried on French aircraft in AJ.37 all-weather anti-radar form only for destroying radar sites. The RAF, however, deploys both versions on its Buccaneers; the AJ.168 is for making stand-off attacks on ground targets, the weapon operator guiding the missile from the picture seen on a monitor inside the aircraft.

73. Sea Skua was developed by British Aerospace as a lightweight sea-skimming anti-ship missile to arm Royal Navy Lynx helicopters at sea, although other aircraft can carry the missile. Entering service in 1981, it has already been in action in the South Atlantic. It can be fired individually or in salvoes and is best suited to attack missile-carrying smaller vessels of fast patrol boat size. Sea Skua's range is 9⅓ miles (15km).

73▼

▲74

74. A French double that is world famous following the recent conflict in the South Atlantic is the Dassault-Breguet Super Etendard with the Aérospatiale AM39 Exocet anti-shipping missile under its starboard wing. Exocet can be air, sea or shore launched. In airborne form, the high-subsonic missile has a range probably up to 43 miles (70km). It travels towards its target at only 7 to 10ft (2 to 3m) above the water, making it a most difficult weapon to detect. (*Air Portraits*)

75. The French 10,000-ton cruiser *Jeanne D'Arc*, with two of its 3.9in guns and six MM38 Exocet launchers clearly in view. Production of the original MM38 ship-launched version of Exocet began in 1973. (French Navy)

76. Known to NATO as 'Backfire', the Soviet Tupolev Tu-22M (sometimes referred to as Tu-26) is the world's only large operational Mach 2 strategic bomber. This 'Backfire' was photographed by a Swedish Air Force interceptor and carries a 'Kitchen' nuclear air-to-surface strategic missile semi-submerged under its fuselage. 'Kitchen' can cruise at more than Mach 2 and has a range of 185 miles (300km). (Swedish Air Force)

77. German Kormoran anti-ship missiles carried by a Marineflieger-configured Panavia Tornado (see **70**).

▼75

76▲ 77▼

▲78 ▼79

80▲

78. In 1964 the US Army first deployed its newly acquired Martin Marietta MGM-31A Pershing selective-range artillery missile, capable of ranges of between 100 and 460 miles (161 to 740km) and carrying a nuclear warhead. In an attempt to increase Pershing's capability, the Pershing 1-A system was developed, introducing a faster erecting launcher, solid-state electronics, and a changeover from tracked to wheeled support vehicles. This photograph shows a Pershing 1-A lifting-off from an erector-launcher comprising a semi-trailer towed by an M-757 tractor.
79. A weapon from Euromissile is the MILAN (Missile d'Infanterie Léger ANti-tank), development of which was completed in 1971. Wire guided, using an optical aiming and infra-red tracking (TCA) system, it can hit a target at only 82ft (25m) distance or at ranges of up to 6,560ft (2,000m). A night firing system has also been developed. MILAN can be vehicle mounted or fired on the ground from a prone position, as illustrated.

80. In 1965 a new phase in aerial warfare was initiated, when Bell flew a prototype of its AH-1 HueyCobra. This was a tandem two-seat attack helicopter, heavily armed and with a fuselage so narrow that it made a difficult target for enemy defences. When it entered production as the AH-1G, it was the world's first purpose-built attack helicopter. With the development of TOW, 92 US Army AH-1Gs were modified to carry eight missiles each as AH-1Qs.
81-84. TOW is an acronym for Tube-launched Optically-tracked Wire-guided, which indicates that it is an anti-tank missile for deployment on the ground and from helicopters. Hughes has produced well over 275,000 TOW missiles, these serving with the armed forces of more than thirty nations. This sequence of photographs shows the first live TOW firing from a helicopter during trials. The helicopter, the Lockheed Cheyenne, did not achieve operational deployment itself.

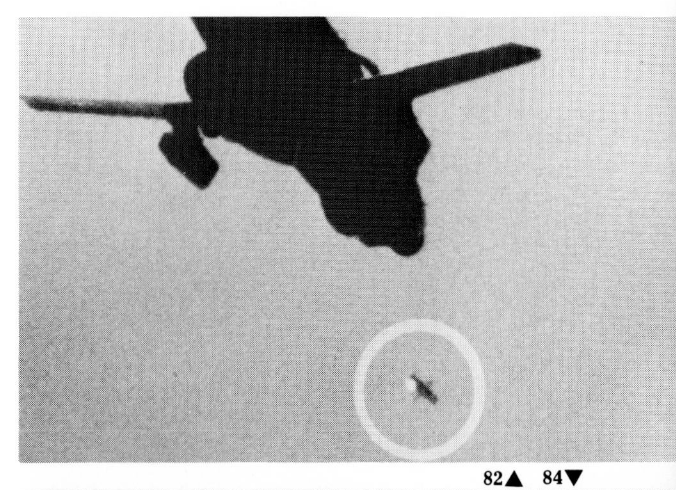

81▲ 83▼

82▲ 84▼

U.S. ARMY
LIFT
HERE

◀85 ▲86

85. This illustration of the US Aerophysics Dart shows clearly the jeep-mounted short-range anti-tank missile on its zero-length launch rail, the joystick and magnification unit. Production was undertaken for the US Army.
86. Pye Ltd. was a British company which developed as a private venture a short-range wire-guided missile known as the Python. Displayed in 1958, Python could be man-portable, as illustrated, the infantryman controlling flight by means of a thumb joystick. Alternatively, six could be carried on a Land Rover or three on a Belgian As 243, in each case making it suitable for anti-tank or fortification roles. The missile was not deployed with the British Army. The illustration of Python does not show the prismatic monoculars used by the controller to sight the far-off missile and target.

▲87 ▼88

89▲

87. Malkara was an anti-tank missile, developed by the Department of Supply in Australia. As is clearly visible from the illustration, it was wire-guided. The missile was adopted for deployment by the British Army, and in 1962 Cyclops Squadron, No. 2 Royal Tank Regiment, began training in its use. Meanwhile, Wharton and Humber developed the Hornet as a three-man carrier, capable of carrying two missiles in launching position and two in store.

88. As a replacement for Malkara, the British Aircraft Corporation developed Swingfire. Also wire-guided, it was designed with two major innovations. Although capable of a range of 13,125ft (4,000m), the solid-propellant rocket motor was developed to give the missile low initial acceleration, making accurate attacks on close-range targets possible. The other innovation is its ability to be launched from around corners in a 90° arc.

89. The illustration shows two Cobra 2000 anti-tank missiles in firing position. One man can launch up to eight missiles from a single control box, no launching rails being required as the missile 'jumps' on take off. Many nations have operated Cobra since 1960 when it entered service (see illustration **2**).

90. The Nord-Aviation Entac (ENgin Téléguidé Anti-Char) was one of the most successful early wire-guided anti-tank missiles, entering service with the French Army in 1957 and thereafter widely exported. Launched from its carrying container, the control system was developed to allow a single man to control up to ten ground-based missiles from one command position, or four missiles could be jeep-mounted, as illustrated.

90▼

91. The SS.10 was another early wire-guided anti-tank missile from Nord, but one which did not achieve Entac's longevity of service despite international use. However, its potential was demonstrated during the Suez crisis of 1956, when Israeli SS.10s were employed with success against Egyptian armoured vehicles.

92. In order to improve upon the SS.10s range, cruising speed and warhead (which was only 11lb; 5kg), and also to supersede Entac, the SS.11 was developed. Several missiles can be grouped on land vehicles or light naval craft, whilst an air-launched version of the missile (AS.11) arms helicopters. An even more recent development is the SS.12 (and similar AS.12), with a 400 per cent increase in warhead weight, at nearly 63lb (28.5kg). Here, an SS.11M and larger SS.12M are deployed on a naval craft.

▲91 ▼92

93. Launch of an SS.12M.
94. Four SS.11 anti-tank missiles on the turret launcher of a tank.

93▲ 94▼

▲95 ▼96

97▲

95. One of the most recent anti-tank missiles is Hot, which is an acronym for High-subsonic Optically-guided Tube-launched. It was developed by Euromissile, an organization formed by companies from France, West Germany and the UK. Hot can be launched from vehicles (Casemate Launcher Vehicle seen here), helicopters or fortifications, and its excellent performance allows targets to be attacked as close and far away as 245ft and 13,125ft (75m and 4,000m) respectively.
96. A Martin Marietta Copperhead about to strike its tank target. Unlike the missiles mentioned previously, Copperhead is fired like a shell from howitzers and seeks the target illuminated by a laser designator. Maximum range is 12½ miles (20km).
97. Another unusual weapon is the Ford Aerospace and Communications MGM-51A Shillelagh, which is launched from the main gun of some combat vehicles that can also fire conventional shells. Unlike Copperhead, Shillelagh is powered by an integral rocket motor and is kept on target by command from the tank. The missile was produced in greater numbers than any other in the West and is deployed on General Sheridan armoured vehicles (as seen) and M60A2 main battle tanks.

▲98

98. An early post-war anti-shipping missile was the Robotbyran 315, work on which began in Sweden in 1949. Development of its pulsejet sustainer motor with four built-in rocket boosters took considerable time, with the result that a firing was not achieved until January 1954. However, by April of the following year the missile was operational on the destroyers *Halland* and *Smaland*. Launched from a long ramp, the missile had a cruising speed of Mach 0.85 and a range of 11½ miles (18km).

99. Currently deployed on *Halland* and *Smaland* is the Saab RB08A, an anti-shipping missile that also forms the armament of coastal defence batteries. Developed from the French CT.20 target drone and carrying a heavy high-explosive warhead, it became operational in the latter 1960s. Here, an RB08A is launched from shore.

100. The Chance Vought SSM-N-8 Regulus I was the US Navy's first operational surface-to-surface missile. It was nuclear armed and powered by a turbojet engine and two solid-propellant boosters. First fired in 1951, 514 were built up to 1959, serving on submarines, guided missile ships, aircraft carriers and from land bases. Here a Regulus I is being loaded into the deck hangar on board the submarine USS *Tunny*.

101. For launching from aircraft carriers, as seen here on board USS *Hancock*, Regulus I was mounted on a jettisonable wheeled trolley. By 1957 ten carriers deployed Regulus. This number was reduced to two by 1960, but the number of Regulus-equipped submarines increased. The phase-out of Regulus I on the five submarines began a few years later. The missile's range was 575 miles (925km). Note the Grumman F-11 Tiger in the foreground.

102. A drone training version of Regulus I was also produced under the Navy designation KDU-1, with recovery and reuse capability. On 8 June 1959 a KDU-1 carried 3,000 letters from the submarine USS *Barbero* to the US Naval Air Station at Jacksonville, recording the first ever official missile mail.

▼99

100▲

101▲ 102▼

▲103 ▼104

103. In 1956 the first launching took place of the Regulus II, a newly developed missile of Mach 2 speed and twice the range of the subsonic Regulus I. Here, a Regulus II is being withdrawn from the hull hangar on board USS *Grayback*, one of only two submarines so equipped before the missile was cancelled in 1958. A number of KD2U-1 Regulus II drones were completed thereafter.
104. Ikara was developed by the Department of Manufacturing Industry, Government of Australia, to provide Australian destroyers with a rapid-reaction anti-submarine missile. The missile has since also been deployed by other navies, including the Royal Navy. Ikara is basically a torpedo-carrying missile, launched as illustrated. During flight, and at the optimum attack position, the torpedo is released by parachute from the airframe, the parachute disconnecting once in the water to allow the torpedo to make its attack.
105. In about 1960 the Soviet Navy first deployed a subsonic ship-launched cruise missile known to NATO as SS-N-2 'Styx'. This still arms the large number of 'Osa' class missile patrol boats in service with the Soviet and many other navies, 'Komar' class patrol boats and larger Indian and Chinese ships. Of aeroplane configuration with a rounded nose, more than 20ft (6m) long and with a high-explosive warhead, 'Styx' claimed the Israeli destroyer *Eilat* as its first victim in 1967. This 'Komar' boat carries two 'Styx' inside the covered launchers. (US Navy)
106. Close-up of the two 'Styx' missiles in their launchers, one either side of the 25mm guns of this 'Komar' class patrol boat. (US Navy)

105▲ 106▼

107. A Honeywell RUR-5A Asroc is fired from the guided missile escort USS *Brooke*. Used by the US Navy since 1961 and thereafter joining other navies, Asroc is basically a missile powered by a rocket motor and with an outer shell containing either a torpedo or nuclear depth charge. At predetermined points during flight, the motor jettisons and later the shell splits to release the torpedo into the water by parachute or drop the depth charge.

108. The Goodyear UUM-44A Subroc was designed to work on similar principles to Asroc, but differs mainly in being submarine launched to attack other submarines. Fired from a torpedo tube, its booster motor propels it out of the water. After the booster has been separated from the nuclear depth bomb warhead, the latter re-enters the water to explode. Subroc became operational with US Navy nuclear attack submarines from 1965, initially of the *Permit* class.

▼107 ▲108

▲109 ▼110

109. Apart from Exocet and 'Styx' already mentioned, there are a number of other anti-shipping missiles in use from naval vessels and coastal defence batteries. Most, such as the Israeli Gabriel and Norwegian Penguin, are rocket powered and have ranges of less than 30-40km. The French/Italian Otomat, first test-fired in 1971, is turbojet powered (plus side-mounted rocket boosters) and has a range of up to 112 miles (180km).

110. A Lockheed UGM-73 Poseidon C3 launched from USS *James Madison* during underwater trials in 1970. This submarine-launched ballistic missile was developed as a follow-on to the Polaris, offering a heavier thermonuclear warhead with multiple independently-targetable re-entry vehicles (MIRV), more accuracy, and yet capable of being carried by modified but existing Polaris submarines. *James Madison* became the first operational Poseidon submarine in 1971.

111▲ 112▼

111. A British Fairey Stooge anti-aircraft missile on its 10ft (3m) launching ramp in about 1947. Conceived as a possible weapon to destroy Japanese suicide planes in the Pacific Theatre before attacks could be pressed on Allied ships, it was not ready for testing before the end of the war. Thereafter, only a small number were built for trials. Radio controlled, it used four integral and four booster rocket motors.

112. The French Marine Française developed a large ship-launched anti-aircraft missile during the 1950s as the Masurca. As can be seen from the illustration, in its Mk 1 development form the upper stage had cruciform triangular foreplanes and larger wings at the tail. Here, a Mk 1 is ready for a trial launching from a land installation.

▲113 114▶

113. Masurca entered production during the early 1960s in a modified Mk 2 form, with redesigned tail surfaces to the upper stage and the foreplanes replaced with long chord and narrow span wings. The booster remained virtually as for the Mk 1. The cruiser *Colbert*, commissioned in 1959, was fitted with a twin launcher, as were the destroyers *Duquesne* and *Suffren* at a later date, each with 48 missiles. The missile to the left is a Latécoère Malafon anti-submarine missile.

114. The Convair SAM-N-7 Terrier was the US Navy's first operational surface-to-air missile. Its development was a spin-off from the 1945 Bumblebee ramjet programme and the missile itself was evolved from the experimental Lark. First test-fired in 1951, Terrier was first deployed in 1956. The missile's range was 12 miles (19.3km). An Advanced Terrier version followed. Here, a Terrier is launched from the battleship USS *Mississippi.*

849

▲115 ▼116

117▲

115. In the 1960s the US Navy deployed a new ship-based anti-aircraft missile as the Tartar (as illustrated), resembling an Advanced Terrier without the booster stage. Range was more than 10 miles (16km), half that of Advanced Terrier but similar to Terrier. Interestingly, at about the same time the Soviet Navy deployed a sea-going version of 'Goa' without booster.
116. In the early 1960s Seaslug, developed by Armstrong Whitworth, was deployed on the first of the Royal Navy's new County class destroyers, HMS *Devonshire.* With a range of 28 miles (45km) and a guidance system proven during trials in Australia, Seaslug offered an effective anti-aircraft system. Here, the missile's four wrap-round boosters jettison. (MoD)

117, 118. British Aerospace Dynamics is the manufacturer of Sea Wolf (**117**) and Sea Dart (**118**). Both missiles made their combat debuts during the Falkland Islands conflict in 1982. Sea Dart is a third generation air defence missile with anti-shipping capability, carried by Royal Navy Type 42 destroyers, HMS *Invincible* and vessels of the Argentine Navy. Sea Wolf is a new rapid-reaction air defence missile, capable of destroying aircraft and anti-shipping missiles, and has even intercepted 4.5in shells. This missile entered service with the Royal Navy in 1979 on HMS *Broadsword*, the first Type 22 frigate. The range of Sea Dart is about 18 miles (30km).

118▼

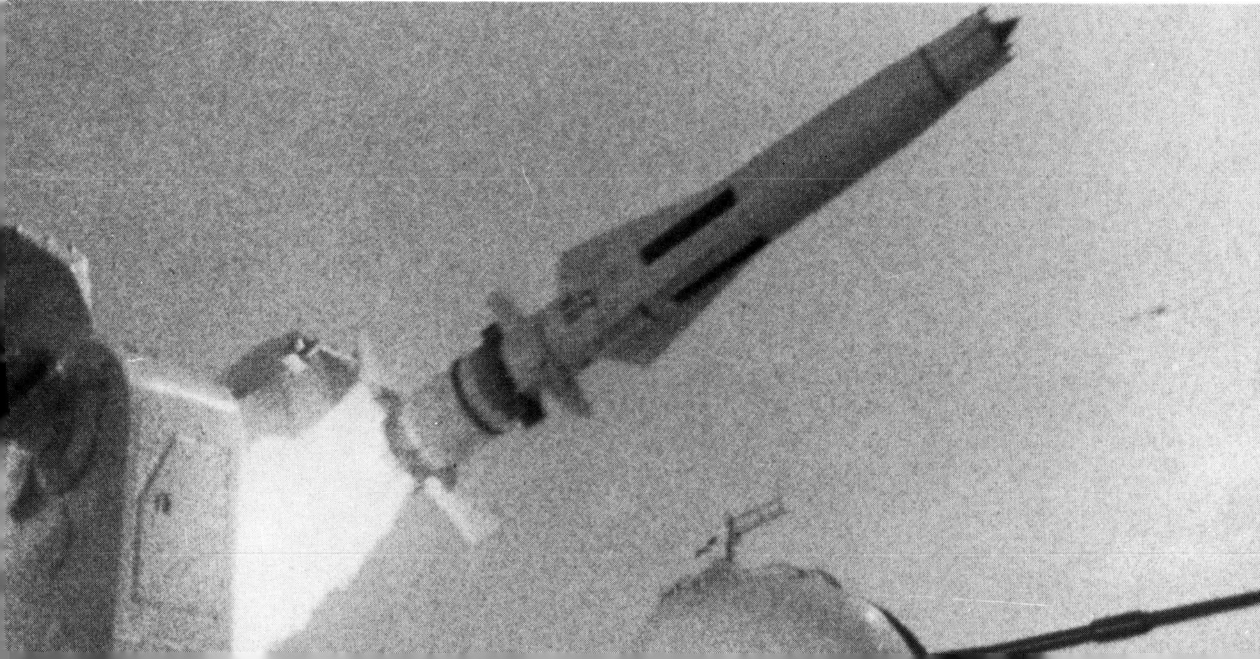

119-121. Development of the Seacat ship-based, close-range, anti-aircraft and surface-to-surface missile was begun by Short Brothers in 1958. Sea trials of what was to become one of the most widely used missiles in the world took place on board the destroyer HMS *Decoy* in 1962, with operational deployment following. Illustration **119** shows a Seacat in its protective handling canister being loaded on HMS *Decoy*; **120,** the missile director and launcher, the director using an optical tracking and radio command guidance system; and **121,** a Seacat being fired from a four-round launcher. A land-based version is also still operated as Tigercat.

◀119

120▲ 121▼

▲122 ▼123

122. During the 1950s, DEFA of France developed the PARCA (Projectile Autopropulsé Radioguidé Contre Avions), seen here on its launcher. Equivalent to Britain's Thunderbird, it was a land-based anti-aircraft missile, powered by a liquid-propellant sustainer motor and with four wrap-round solid-propellant boosters. Guidance was by radio command. When PARCA entered service with the French Army, it became France's first operational missile.

123. As a replacement for Nike Ajax, Western Electric developed the M-6 (later MIM-14B) Nike Hercules. With a speed of Mach 3.65, a range of 87 miles (140km), and alternative high-explosive or nuclear warheads, it was far more formidable. Here, Nike Hercules missiles on mobile launchers are readied during the national alert caused by the Cuban Crisis in 1962. (US Air Force)

124. In September 1958, Battery B, 1st Missile Bn, 562nd Artillery Regiment, became the first Nike Hercules unit on the US East Coast, protecting Washington and Baltimore. (US Army)

U.S. ARMY
204

▲125

▲127 ▼128

125. Boeing's mighty IM-99 Bomarc (later USAF designated CIM-10) was the first long-range anti-aircraft missile to go into service with any nation. The IM-99A initial version had a range of 250 miles (402km). Prototypes were first launched in 1952. Power was provided by two ramjet sustainer engines under the body and a rocket booster in the tail. The warhead was nuclear. Here asbestos-suited crew prepare a Bomarc at Patrick AFB.
126. The IM-99B or CIM-10B Super Bomarc became operational in 1961, thereafter serving at six US and two Canadian bases. Of 700 Bomarcs built, only 188 were of this version, which had a range of 440 miles (708km). All Bomarcs received radio command guidance from SAGE (Semi-Automatic Ground Environment), the USAF's electronic warning and air defence system. Here, an IM-99B thunders from its launcher on booster power.
127. The Bristol Bloodhound was the RAF's first surface-to-air missile, entering service in 1958. Illustrated is a Bloodhound Mk 1 arriving at the Fire Unit of RAF North Coates, Lincolnshire, in 1960. The Mk 2 version followed the Mk 1, similarly powered by two Bristol Thor ramjet sustainers and four jettisonable wrap-round boosters, and with a 50-mile (80km) range. Other users have included Australia, Singapore, Sweden and Switzerland.
128. The English Electric Thunderbird looked similar to the Bloodhound, but was smaller, entirely mobile and with a sustainer motor inside the body. It went first to Nos. 36 and 37 Anti-Aircraft Regiments of the British Army in 1960, thereafter also being exported. Here, a Thunderbird is undergoing low temperature tests.

▲129 130▼

131▶

129. In 1960 the US Army began the deployment of the new Raytheon MIM-23A Hawk (Homing-All-the-Way Killer), an anti-aircraft missile capable of destroying targets flying at low and medium altitudes. These, and the missiles received by the USMC, were deployed in many parts of the world, including Vietnam. Many other nations have acquired Hawk, including France, Italy, the Netherlands and West Germany. The illustration shows a Sikorsky S-56 Mojave lifting Hawks belonging to the US Army.
130. A few years after deployment of Hawk began, development of the MIM-23B Improved Hawk started. As for Hawk, Improved Hawks can be mounted on a three-round trailer launcher or on a tracked vehicle. Improved Hawk introduced an improved rocket motor, larger high-explosive warhead, a solid-state guidance package and other refinements.
131. The Ford Aerospace MIM-72A Chaparral battlefield weapon system was developed from the mid-1960s to equip US Army air defence battalions then forming. Based on the Sidewinder 1C missile, modified to the new role, four are carried on launchers on the M730 tracked vehicle (as seen), on a trailer or less mobile platform. Two Chaparral batteries are backed by two batteries of Vulcan guns.

US ARMY
12GJ 06

132. Rapier was developed in the 1960s as a quick-reaction missile system to combat aircraft flying at below 10,000ft (3,660m). Its mobility was demonstrated during the recent action in the South Atlantic, when Rapiers were deployed rapidly in key positions once landing forces were ashore. The illustration shows the four-missile launcher with surveillance radar and the optical tracker forward of this. A radar tracker can replace the optical tracker under the 'blindfire' system, for automatic and night/poor visibility interception. Rapier is used by several nations, including Australia, Iran, Kenya and Zambia. It has also been bought by the United States for use in the protection of their air bases in the UK.

133. Roland, like Chaparral, is a battlefield close-range anti-aircraft missile system, suitable against targets flying at less than Mach 1.3. Developed by Euromissile and first deployed by France and Germany in 1977, the Roland I system has two tube-launched missiles on a combat vehicle and uses an optical tracking system. Roland II is the all-weather derivative with a tracking radar, as seen in the illustration on a German Army SPZ Marder vehicle. Other nations deploying Roland include the USA, which licence-builds the system as the US Roland.

134. The first of the one-man portable, shoulder-fired, close-range anti-aircraft missiles was the FIM-43A Redeye, developed during the early 1960s by Convair (General Dynamics). Used by the US Army, USMC and armies of other nations, the 4ft (1.22m) missile is launched from the tube by a boost charge, allowing the missile to fly away from the operator before the sustainer rocket ignites. After being optically aimed, an infra-red terminal homing guidance system takes over.